International Webinar on Ships' Life-Cycle 2021

Online
31 March 2021

ISBN: 978-1-7138-8305-0

Printed from e-media with permission by:

Curran Associates, Inc.
57 Morehouse Lane
Red Hook, NY 12571

Some format issues inherent in the e-media version may also appear in this print version.

Printed with permission by Curran Associates, Inc. (2024)

For permission requests, please contact The Royal Institution of Naval Architects at the address below.

The Royal Institution of Naval Architects
8-9 Northumberland Street
London, WC2N 5DA
United Kingdom

Phone: 020 7235 4622
Fax: 020 7259 5912

publications@rina.org.uk

Additional copies of this publication are available from:

Curran Associates, Inc.
57 Morehouse Lane
Red Hook, NY 12571 USA
Phone: 845-758-0400
Fax: 845-758-2633
Email: curran@proceedings.com
Web: www.proceedings.com

International Webinar

Ships' Life-Cycle

31 March 2021

PROCEEDINGS

THE ROYAL INSTITUTION OF NAVAL ARCHITECTS
8-9 Northumberland Street, London, WC2N 5DA, Telephone: +44 (0)20 7235 4622

The Royal Institution of
Naval Architects

INTERNATIONAL WEBINAR

Ships' Life-Cycle

31 March 2021

Papers published in these conference Proceedings are peer reviewed.

THE ROYAL INSTITUTION OF NAVAL ARCHITECTS
8-9 Northumberland Street
London WC2N 5DA

Telephone: 020 7235 4622
Fax: 020 7259 5912

CONTENTS

A MORE USER FRIENDLY APPROACH FOR USING CAD AND PLM INTEGRATIONS

R Penas and **R Perez-Fernandez,** SENER Ingeniería y Sistemas S.A., Spain

SUMMARY

In recent years, the integration capability between CAD and PLM systems has become a key functionality to ensure productivity in ship design and construction, to guarantee an optimal quality, reduce design errors, enable the appropriate digitalization processes and allow consuming CAD generated data along the whole ship life cycle.
However, many different approaches have been adopted by the largest shipyards to implement this feature, from the simple export/import based processes, through gateway functionalities, and up to a full bidirectional real-time integration.
But now, we are presenting a step beyond, which is the interoperability between both worlds: CAD and PLM. The current paper describes some of the most recent capabilities being introduced, by means of having a really user friendly handing of data between them.

NOMENCLATURE

CAD	Computer Aided Design
ECN	Engineering Change Notice
PLM	Product Lifecycle Management

1. INTRODUCTION

During the last decades, a tight and efficient integration between the CAD and the PLM systems used in complex shipbuilding products, definition, manufacturing and operation has become a more and more important matter, in order to ensure design efficiency, avoid design errors, prevent wasting time in using wrong data or, simply, to provide a common environment in which different organizations and partners working in the same shipbuilding design programs must share their data and follow common procedures.

Some software suppliers provide both tools, CAD and PLM, and a high level of integration between both worlds but, in many cases, using a single software supplier is not feasible due to different reasons (Perez and Penas, 2015):

- The CAD tool that a company uses is strongly implemented in one of the program partners, while the PLM system to share the data is provided by a different supplier.
- The optimal and more productive CAD tool comes from a supplier that either does not provide a PLM system or it does not fit the expected requirements.

Whatever the reason, transferring the data between CAD and PLM is a strong requirement, and depending on the integration level reached, the efficiency in data handling can be higher or lower.

2. DATABASE VS. FILE BASED CAD SYSTEMS

When we talk about CAD and PLM integration, we must distinguish between CAD systems that use individual files to store the data (files that must be put under a PLM system control to be properly managed) and CAD applications that use their own database as a data repository.

The first one is a classic mechanical CAD to PLM integration approach, while the second is actually a proper case of systems integration, as both applications make use of their own database and both of them have to be maintained linked and synchronized, in order to guarantee data consistency and, as well, to ensure that the principle of the PLM as the "single point of truth" is maintained.

In this paper, we will be centred in the database CAD systems approach.

3. INTEGRATION VS. GATEWAY

3.1 WHAT CAN BE UNDERSTOOD AS AN INTEGRATION BETWEEN TWO APPLICATIONS

Two applications can be communicated in many different ways, and communication level can be more tight or loose.

In any case, we consider the existing communication level between two applications can only be considered an "integration" if both the following conditions are accomplished:

- Updating the data in one of the systems makes that the data will also be updated in the other system without any further user action. Saving data is performed only once and will be updated in both systems, with any additional user task.
- The data update is performed in both systems in real time, without any significant time delay.

According to this, systems in which the user first saves the data in one system but then some further manual action has to be performed to update the data in the other, cannot be considered as an integration, but just a gateway.

Likewise, systems in which the user saves the data only once in one of the systems, but the other one is updated by a delayed in time automatic task (e.g. by some automatic nightly executed trigger that "publishes" all modified data to the other system), is neither an integration.

3.2 CAD AND PLM LEVELS OF INTEGRATION

As a general view, the level of integration that may exist between a CAD and a PLM system can be classified into the following levels:

- Level 1. CAD data is manually exported and imported into PLM system, using generic export/import procedures. Unidirectional CAD→PLM. This cannot be considered properly as an integration or a gateway, but just an export/import procedure. This is the approach normally used when no direct communication capability exists between previous systems.

- Level 2. CAD data exported and imported into PLM by some automatic or manual on-demand ad-hoc dedicated procedures invoked from either the CAD or PLM tools. Unidirectional CAD→PLM. This ensures a better data management quality than level 1 approach, but still cannot be considered yet as an integration, and is normally called a Gateway.

- Level 3. Similar to Level 2 approach, but adding bidirectional capability CAD↔PLM, by enabling some PLM to CAD data transfer, to apply to CAD system some changes that have been produced in PLM. Like in CAD to PLM data transfer, PLM to CAD communication can be launched either by some automatic tasks (normally invoked as a result of certain events, like approval processes) or by manual on-demand ones.

- Level 4. Having the bidirectional capabilities of Level 3 approach, but adding a sort of interoperability capabilities. Enable direct access to PLM system embedded into CAD tool that allow performing actions in PLM on entities selected in CAD, or vice-versa, performing CAD actions on entities that have been identified and selected in PLM and we have loaded in the CAD scene for working on them. This paper will be mostly dedicated to describe this interoperability functionality.

- Level 5. Considering all the capabilities available in a level 4 approach, we add on them the function of saving data in PLM system at the moment we do it in CAD, in real time, without having to dedicate a subsequent either automatic or manual on-demand task to "publish" the CAD data into PLM.

Taking into account the criteria described in the former section abut what can be considered as a "full integration" capability between two different applications, it is easy to conclude that only a Level 5 approach can actually be accepted as a real "integration".

4. CAD-PLM INTEROPERABILITY ELEMENTS AND UTILITIES

4.1 CAD PRODUCT TREE

A full tree available access to the whole CAD model. In FORAN system, it is called FORAN Product Tree or, FPTree. From this utility, we can:

- Load model data in the 3D scene.

- Access entity properties in PLM manager.

- Display status of entities in PLM (created, published, pending to be published, synchronized, locked …).

- In the case of a project in a series of ships, the applicability range of the entities (Perez and Penas, 2016).

- Launch entity publishing to update data in PLM for it.

- Launch an external web browser to access entity data in PLM.

- Drag & drop the entities to access data in PLM inside an internal browser.

- Drag & drop the entities into the scene.

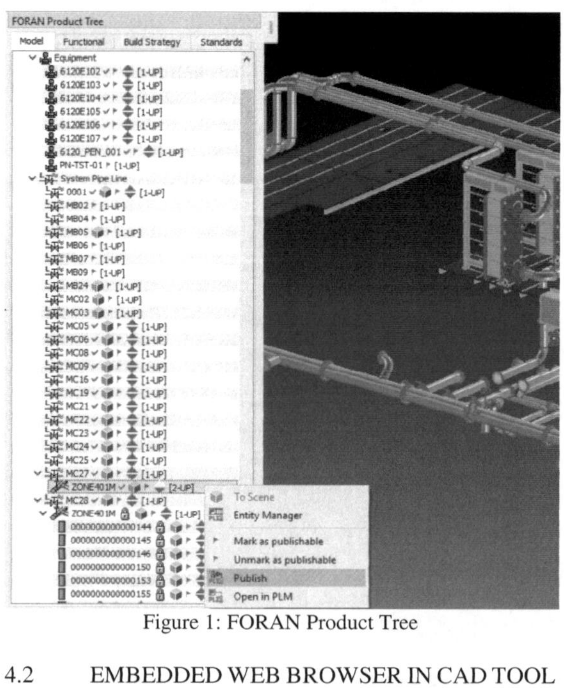

Figure 1: FORAN Product Tree

4.2 EMBEDDED WEB BROWSER IN CAD TOOL WITH ACCESS TO PLM

A web browser with access to PLM server via web is available inside CAD modules, to allow access to the PLM data for the selected CAD entities, as well as to interact with CAD scene by selecting objects navigating into the CAD scene.

The browser allows the access to two possible servers only:

- The PLM server connection. It is the main access allowed. It is previously configured and the user will have access to it. This is to avoid accessing other PLM servers with different data than those mapped than represent the CAD model.

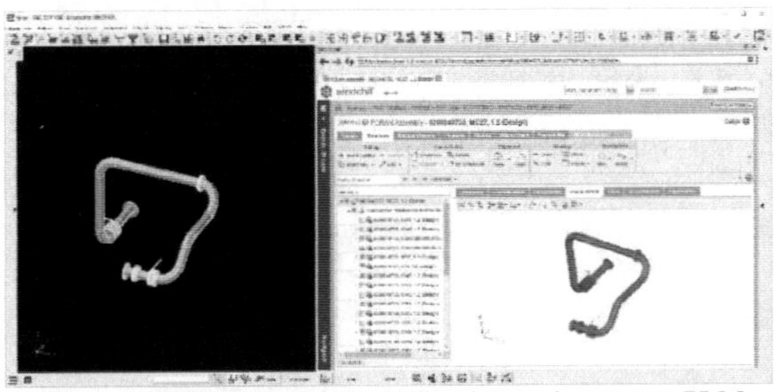

Figure 2: FORAN module including internal access to PLM (Windchill) server

- The FORAN FPLM Web Server. In the case of having synchronization (events connection from PLM to FORAN) enabled, the internal web browser will also be allowed to access that FORAN web server, in order to visualize synchronization results directly from that server.

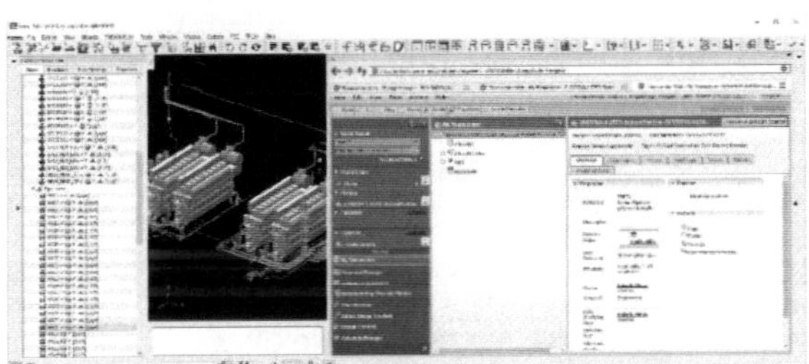

Figure 3: FORAN module including internal access to PLM (Teamcenter) server

4.3 INTEGRATED UTILITIES IN THE CAD SESSION

The different elements appear by default, in the following places:

- The FORAN Product Tree at the left.
- The 3D scene in the middle.
- The internal PLM access embedded browser at the right, although it can be undocked and located anywhere in the screen.

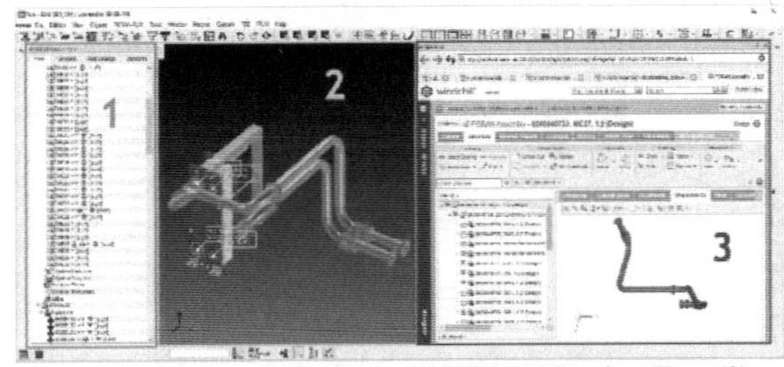

Figure 4: Integrated elements: (1) FORAN Product Tree (2). FORAN 3D scene. (3) Internal web browser with direct access to PLM server (example with Windchill)

5. CAD-PLM INTEROPERABILITY ACTIONS

5.1 DATA FLOW

The following diagram shows the different interoperability actions that could take place.

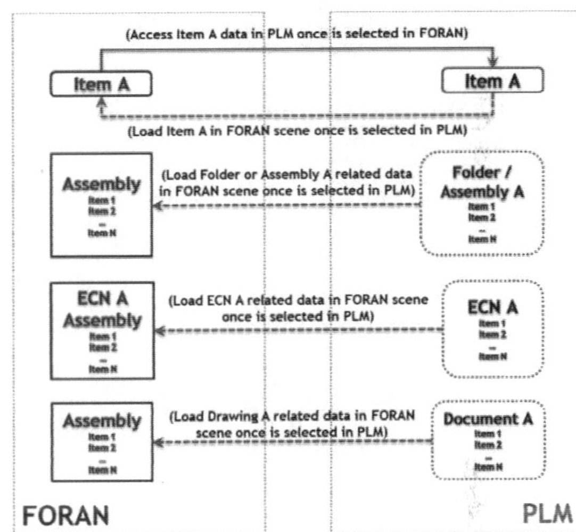

Figure 5: CAD & PLM interoperability flow

5.2 DESCRIPTION OF RELATED ACTIONS

5.2 (a) Direct access to data in PLM for an entity identified in CAD

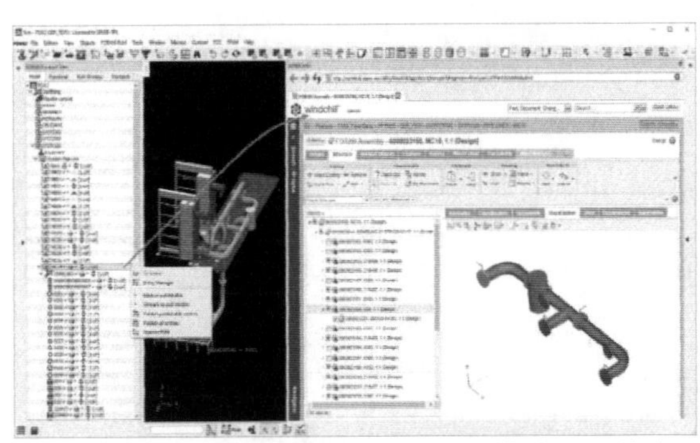

Figure 6: Access PLM data for a CAD identified assembly by drag & drop from the FORAN Product Tree

Identify a CAD entity either displayed in the 3D scene or selected in the CAD Product Tree, a directly open that entity properties page in PLM system. The entity can be either a single item or an assembly. Data is accessed in PLM browser window by performing a drag & drop action from CAD.

5.2 (b) Access both CAD and PLM data from a single click in CAD scene

Select an item in the CAD scene and the system, at the same time, will show the CAD data for it in the CAD properties page, as well as access to PLM properties page for that item.

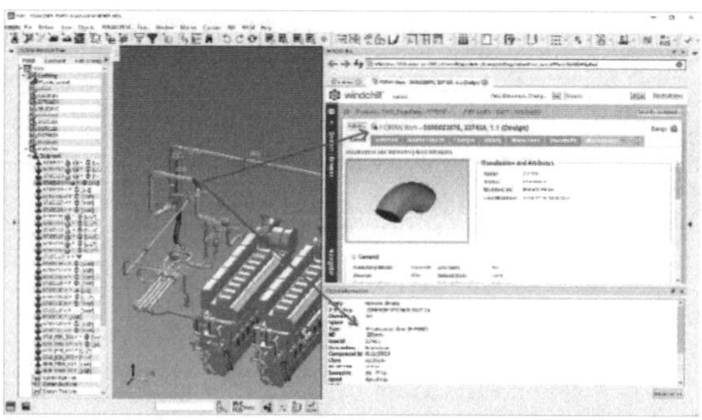

Figure 7: Select an item in CAD scene and directly show simultaneously the item data in CAD and PLM

5.2 (c) Load data in CAD 3D scene by selecting items in PLM

Identify an entity in PLM browser window and load it into CAD scene by drag & drop from the page URL or from any hyperlink.

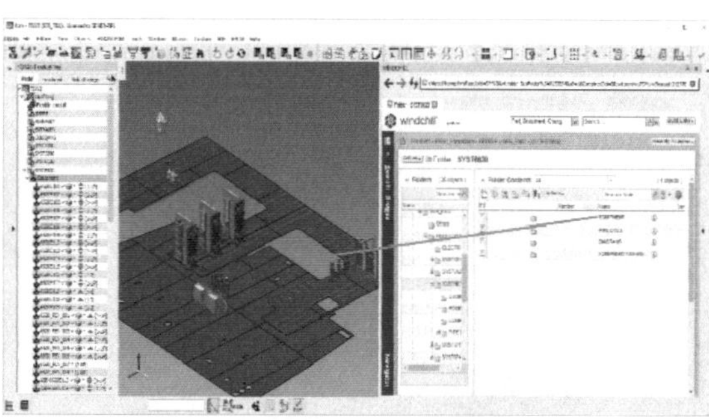

Figure 8: Load a PLM (*Windchill*) folder contained items into CAD scene, by drag & drop from a folder hyperlink

5.2 (d) Load the related PLM ECN data in CAD 3D scene by selecting the ECN in PLM

Identify an ECN in PLM window and load its related items into CAD scene by drag & drop either from the ECN page URL or from any hyperlink to it. ECN related entities can be affected, impacted or solution ones.

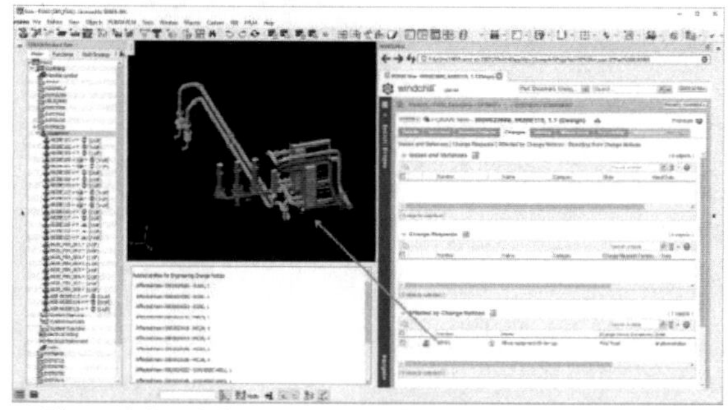

Figure 9: Load the PLM ECN related data into CAD scene

5.2 (e) Load the related PLM document or drawing described data in CAD 3D scene by selecting the document in PLM

Identify a document or a drawing in PLM window and load its described items into CAD scene by drag & drop from the document page URL or a hyperlink to it.

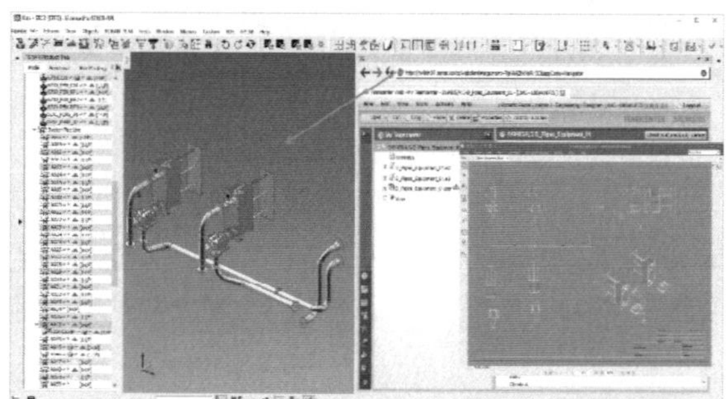

Figure 10: Load a PLM (*Teamcenter*) drawing related data in CAD 3D scene by drag & drop from PLM drawing URL

6. REACHING INTEGRATION LEVEL 5

The maximum level of integration can be reached as soon as it is possible to launch publishing to PLM when the data is stored in CAD repository DB. The user could store the modification changes performed and, at the same time, the system will launch the publishing of those changes into PLM system.

In order to reduce the returning times of CAD module control, when a user stores the data, publishing to PLM could be launched in a separate thread, so that publish results could then be communicated to CAD database by invoking the appropriate CAD API functions for this purpose. Working that way, the user would recover the session control, while a task is triggered to update the changes to PLM.

In addition, the existing synchronization tasks (those launched from PLM to CAD) can be produced in complete real time, by launching synchronization tasks associated to certain event triggers, like maturity changes performed by a PLM workflow.

7. CONCLUSIONS

The experience in the usage of these new interoperability tools between CAD and PLM shows an extremely positive response on users. Some of the clear advantages and benefits are listed below:

- Increase time saving by avoiding working with the wrong data.
- Increase time saving by allowing selecting the data only once (in the usual single point of truth: the PLM system) and perform the data loading in CAD with the items already identified in the PLM, instead of having to locate the data again in the CAD tool.
- Allow data loading in CAD according to item selections performed in the PLM:
  - Data included in a PLM folder.
  - A set of items included in a product structure created and updated in the PLM, not necessarily linked to any criteria in CAD.
  - Data related to an Engineering Change Notice (ECN).
  - Set of items described by a particular drawing or document whose relationships have been defined in the PLM, not in the CAD.
  - Immediate and direct access to the PLM data just published from the CAD, without having to search the published data in the PLM.
  - Simultaneous access to CAD and PLM data once an item is selected from the CAD scene.
  - Perform any PLM action from the internal browser, without having to set an external session with PLM.

User friendliness has been dramatically increased. The accumulated experiences show that, once users got used to these new tools, they do not want to go back to previous software releases missing them.

In addition, productivity and ease of use go together when using these new capabilities.

As well, these new tools are and could be available with any PLM system including an appropriate web access. They are currently fully available from FORAN system with *Windchill* and *Teamcenter* PLM systems.

8. REFERENCES

[1]. PENAS, R., SANCHEZ, L., GOMEZ, A. and PASTOR, L. (2009). *A neutral framework for the integration of CAD in product model lifecycle systems*. ICCAS, Shanghai, Sep. 2009.

[2]. PENAS, R. (2013). *An Advanced Step in Integrating Database-Controlled Shipbuilding CAD with PLM Systems*. ICCAS, Busan, Sep. 2013.

[3]. PEREZ, R. and PENAS, R. (2015). *Integration between shipbuilding CAD Systems and a generic PLM tool in naval projects*. Computer Science and Applications, Vol. 2, N. 5, pp. 181-191.

[4]. PEREZ, R. and PENAS, R. (2016). *How to manage sister ships on a shipbuilding CAD tool. The applicability concept*. Brodogradnja, Vol. 67, N. 4, pp. 49-59.

9. AUTHORS BIOGRAPHY

Roberto Penas, Master Naval Architect, is the technical leader in integrating CAD and PLM systems in *SENER Ingeniería y Sistemas S.A.*, with more than 30 years of experience in architecture, development and implementation of CAD and PLM integrated solutions.

Rodrigo Perez, Ph.D. is the head of *Naval Shipbuilding Business Development* in *SENER Ingeniería y Sistemas S.A.* He is also professor in several subjects at the *Marine Engineering School* of the *Technical University of Madrid*.

THE INVENTORYING OF HAZARDOUS MATERIALS, RESPONSIBLE RECYCLING PRACTICE & REDUCING HAZARDOUS MATERIALS, DURING A SHIP'S LIFECYCLE

J Ren Di, CTI-MARITEC Environmental Services, Singapore
A Lougridis, CTI-MARITEC, Greece

INTRODUCTION

Hazardous Materials that may exist onboard ships (as part of their fixed structure or machinery), are listed within the Hong Kong Convention for the safe and environmentally sound recycling of ships 2009 and associated circulars, also within the Annexes of Regulation (EU) 1257/2013 of the European Parliament-Council on ship recycling. Also listed are threshold values i.e., the concentration in homogeneous materials that make a material hazardous.

The regulations require that (a) such materials are identified and inventoried (b) kept track of during the lifetime of the ship, (c) asbestos found onboard ships built after 2002 be removed within a stipulated period and (d) when a ship is recycled, the recycling process is done in a safe and environmentally considerate manner, all hazardous materials are taken off systematically and disposed so as to not endanger living organisms or cause environmental pollution. In this paper we will discuss the above processes.

1. DEFINITION - HAZARDOUS MATERIAL

A hazardous material is any item or agent (biological, chemical, radiological, and/or physical), which has the potential to cause harm to humans, animals, or the environment, either by itself or through interaction with other factors.

2. TYPES OF HAZARDOUS MATERIAL

Types of hazardous materials listed in the relevant regulations are mentioned below for reference

Asbestos
Polychlorinated Biphenyls
Ozone Depleting Substances
Antifouling paints containing Organotin
Perfluorooctane sulfonic Acid
Cadmium and Cadmium Compounds
Hexavalent Chromium and it's Compounds
Lead and Lead Compounds
Mercury and Mercury Compounds
Polybrominated Biphenyl (PBBs)
Polybrominated Diphenyl Ethers (PBDEs)
Polychlorinated Naphthalenes (> 3 chlorine atoms)
Radioactive substances
Certain Shortchain Chlorinated Paraffins
Brominated Flame Retardant (HBCDD)

Table 1 – Types of Hazardous Materials

3. LOCATIONS OF HAZARDOUS MATERIALS ON SHIPS

Hazardous materials are generally found in the locations stated within the following table. The locations stated are not exhaustive and a proper study of the ships plans, drawings, machinery particulars, paint systems, structural material and insulation and other documents is generally to be made before an Inventory of Hazardous Materials is commenced.

Asbestos	• Lagging material for piping flanges, fuel pipes, exhaust pipes of engines • Boiler & Incinerator insulation/manhole packings • Fuel tank lagging • Fire door insulation and accommodation insulation • Sealing putty • Brake linings • Propeller shafting packings or synthetic stern tubes
Polychlorinated Biphenyls	• Transformer Oils • Electric Cable Sheathing • Rubber Isolation Mounts • Fluorescent light ballast
Ozone Depleting Substances	• Refrigerants for Freezers & Air Conditioning • Blowing agent for insulation of LNG tanks • Extinguishing agents
Organotin compounds	• Tin based anti-fouling paints– not permitted, but some ships have the old tin-based paint overcoated by permitted paints.
Cadmium	• Plating films & Bearings
Lead	• Batteries • Certain primers • Insulation • Solder
Mercury	• Liquid level indicators • Thermometers • Mercury lamps • Fire detectors • Gyro compasses
Radioactive substances	• Smoke detectors • Radioactive paints - radionuclides • Level gauges

Table 2 – Some Locations of Hazardous Materials

4. INVENTORY OF HAZARDOUS MATERIALS (IHM) – SHIPS IN OPERATION

4.1 THE IHM PROCESS (PART I)

- The Owner/Operator selects a certified (preferably by an IACS Class or major Flag State) IHM service provider to perform the IHM
- The service provider will ask for (amongst other things) the following documents - General Arrangement Plan, Capacity Plan, Accommodation Plan, Machinery Arrangement Plan, Manufacturers Manuals for selected machinery, Fire and Life Saving Plan, Insulation Plan; Material Specification for deck-heads, side panels, deck flooring, doors, insulation material used, joiner material; Piping Insulation, AFS Certificate, Ozone depleting Substances Log, Spare Parts and Tools List, Repair and modification history, Asbestos Free Declarations from builders or shipyards (it may be noted though that in several cases asbestos has been found on ships with an Asbestos Free Declaration) , etc..
- Based on a study of the above the service provider will draw up a Visual / Sampling Check Plan (VSCP) and a 'Location Diagram' (LD) for use by the IHM certified surveyor(s), who will be performing the survey.
- The surveyor(s) will go through the VSCP and LD and perform an onboard survey during which they will examine the areas that may contain hazardous materials, and/or take physical samples from such locations as deemed necessary. About 80 samples are generally taken on an average.

Fig 1 – CTI-MARITEC hazmat expert performing an IHM survey on a ship in service

- The surveyor or survey team will then carefully mark and label the samples and courier or hand carry the samples to a certified laboratory
- The laboratory will analyse the samples according to the regulatory specifications and issue its report
- The laboratory results will be taken into account by the service provider when drawing up their report
- The service provider will issue their Final Report on the presence and location of any HMs onboard
- The Final Report will be sent for approval to the vessel's Class or Flag. A Class or Flag surveyor will counter verify the report during the next visit to the vessel whereafter a Statement of Compliance or Certificate will be issued to the ship.

4.2 MAINTAINENCE OF THE IHM

Regulations require that the Inventory of Hazardous Materials is maintained up to date throughout the operational lifetime of the ship. This means that the IHM process will be ongoing, in that any changes made to the ship's structure or material connected thereto, or to the ship's equipment, that may cause a change to the condition found during the original IHM survey, shall be duly recorded.

Replacement parts or material, including stores containing hazardous materials, will need to have their MD (Material Declaration) and SDOC (Supplier's Declaration of Conformance) carefully read to determine if any item supplied contains material defined as hazardous and the IHM documentation and records updated accordingly when such part or material is used. This is rather laborious and several software sets are available to help.

4.2(a) STEPS FOR PROPER IHM MAINTENANCE

- A person in the Company has to be Designated as Person in Charge of IHM Maintenance
- The Company must identify High-Risk items suspected to contain hazmat (in line with indicative list in IMO'MEPC 269(68)) and follow-up with suppliers for MDs and SDoCs of such items
- Whenever new equipment is fitted, structural alterations made, or paint systems altered, that may cause a change in the status of hazardous materials onboard, MDs and SDOCs need to be requested from the firms concerned and items associated with hazardous material (either added or removed) identified.
- The IHM shall then be updated when hazardous material is added or removed from structure or

equipment, unless exactly similar parts/equipment are used.

- The Designated Person shall check the updated IHM for correctness and liaise with the ship's Master for the latest revision to be held onboard.
- The updated IHM to be endorsed & verified by Class at the next certificate or Statement of Compliance renewal date.

4.2(b) COMMON PROBLEMS FACED BY IHM MAINTENANCE STAFF

- No Material Declarations (MD) and Supplier's Declaration of Conformance (SDOC) received along with the supplied material, requiring sending constant reminders and follow-up by the shipping company's staff.
- Incorrect or partially filled-up Material Declarations (MD) and Supplier's Declaration of Conformance (SDOC) from the Supplier of Material to ships
- Receipt of MD, SDOC not in a format prescribed by the regulations.
- Suppliers providing only asbestos free declarations and not information on other hazardous materials.
- Ignorance on the part of suppliers regarding the regulatory requirements, requiring them to be educated. (It must be understood that most suppliers are not manufacturers and rely on the manufacturer to give them proper MDs and SDOCs. The Manufacturer in turn relies on upstream suppliers of raw materials who are often quite unaware of the regulatory requirements for Inventory of Hazardous Materials applicable to shipping or regulations made by the International Maritime Organization)

4.3 IHM PART II AND III (AT TIME OF RECYCLING)

IHM Part II lists operationally generated wastes that may contain hazardous or pollutive material that will be onboard at the time of a ship proceeding for recycling.

Part III lists consumable stores and spare parts and liquids and gases sealed within ship's machinery that may contain hazardous or pollutive material

4.4 ASBESTOS FOUND DURING IHM SURVEYS **(see also Table 3 in section 6.3)**

By SOLAS regulations, ships built after 2011 shall have no asbestos onboard. Ships built between 2002 and 2011 may have asbestos fitted but only in certain specified machinery. If during an IHM survey, asbestos is found onboard a ship built after 2002 (except where permitted in certain machinery on ships built before 2011), then regulations required that it be removed within a period of 3 years and replaced with a non-asbestos equivalent. This replacement must not be attempted by ship's crew but must be done by trained and certified professionals.

Fig 2 – CTI-MARITEC expert prepares for asbestos removal on a ship in service.

4.5 IHM PROCESS – NEW BUILDINGS

The Ship Building Yard draws up an Inventory of Hazardous Material by gathering information from Suppliers and Subcontractors (through Material Declarations and Suppliers Declaration of Conformance) on materials used during construction and outfitting, then listing within it, hazardous materials (as identified in the IHM Regulations) onboard.

The Ship Building Yard has the Inventory of Hazardous Material certified after being approved by Class or Flag. The Yard hands over the Inventory of Hazardous Material to Owner at time of delivery

The Ship Owner maintains the Inventory of Hazardous Material up to date throughout the lifetime of the ship, recording additions or removal of hazardous material. In case of sale of the ship the IHM documents must be handed over to the new Owner.

5. RESPONSIBLE (GREEN) SHIP RECYCLING

5.1 BACKGROUND

Ships of all sizes, types and trades eventually reach the end of their economical lives due to wear and tear, outdated equipment and design, mechanical problems, intensified maintenance that is required, also the high

costs of meeting the requirements of new rules and regulations.
So, the Owner has no choice but to take them out of service and either lay them up or recycle them. However, recycling regulations presently in force, or coming into force, require that the Owner recycle these ships in a responsible manner.

But often the Owners are faced today with a bewildering number of rules, regulations, guidelines, forthcoming rules, press and magazine articles, webinar or conference handouts and inter-company information exchanges on what constitutes responsible (or "GREEN") Ship Recycling.

Fig 3 -The days work commences on a ship being recycled under CTI-MARITEC supervision.

5.2 SO WHAT DOES 'GREEN' REALLY MEAN

Although the 'green' connotation may lead us to believe that it deals mainly with the protection of the environment, it actually means that: -

- Safety, Health and Dignity of the human workforce involved in the recycling process remains paramount and work practices and conditions of employment ensure this.
- The ship itself is dismantled in a manner which endeavours to cause no pollution of the surrounding waters, no pollution to the land by permeation of the soil and no pollution of the atmosphere.
- Hazardous materials onboard the ship are identified, demarcated and carefully removed prior to, or during the recycling process, then disposed of in a manner which would render them harmless, or sealed off to completely prevent their release to the environment or exposure to living creatures in the future.

5.3 APPLICABLE REGULATIONS

In the main, recycling regulations today are dictated by –
>The Hong Kong International Convention for Safe and Environmentally Sound Recycling of Ships 2009 (not yet in force, but voluntarily met by many parties, also in cases, in the process of being incorporated into National Laws, such as the Indian Recycling Act 2019).
>The European Union Regulations on Ship Recycling EU 1257/2013.
Note 1: The EU requires that EU flag ships are recycled only at EU 'white listed' yards/facilities and has so far 'white listed' 43 yards – 34 in the EU, 8 in Turkey and 1 in the USA. (A reminder: Brexit means that 3 UK yards will exit the EU list, leaving 40 yards listed)
Note 2: EU requirements are that ships are recycled by facilities operating from 'built structures' such as quays, pontoons, floating docks or dry docks, slipways, ramps, ship-lifts, enclosed canals and the like, so that there is little chance of any environmental damage to surrounding waters. Facilities using the 'landing' method requires them to take extra precautions to ensure impermeable floors under all operating worksites.

5.4 OWNERS RESPONSIBILITIES

So then, how do Ship Owners, with a high standard of Corporate Social Responsibility, ensure that their ships to be recycled are done so in a "green" manner, regardless of the geographical location of the yard or facility?

Well, they need to: -
Select a recycling yard that is certified (ideally by an IACS member) for compliance with the Hong Kong Convention (which although not yet ratified still remains a recognized reference standard); or a yard that has been 'white listed' by the EU. Also, that the yard holds an official Document of Authorization for Recycling (DASR).

Preferably have the yard pre-audited, to ensure that-

- The yard layout and operational features actually comply with the requirements of the certificate and authorization it possesses and that the yard fully meets the requirements of its own approved Ship Recycling Facility Plan.
- The Yard's Senior Management/ HSE Personnel are sincere and knowledgeable, aware of all foreseeable risks and have in place a risk management programme.
- The yard has a properly trained and fit workforce who are equipped with all the necessary Personal Protective Equipment required for the jobs they will undertake. The Yard has in place measures to ensure the health of the work force including regular medical tests, especially of workers involved in hazardous material handling.
- The Yard has all the necessary equipment in good condition (certified where necessary – example certified cranes) to perform proper and safe recycling activities.
- The Yard is well sited so as not to be a hazard to surrounding communities and reasonably

protected from possible damaging effects of natural forces.

- The Yard has in place measures to protect the environment from any type of pollution – be it airborne, liquid or solids and to monitor their own performance in doing so.
- The Yard has the means to carefully identify hazardous material based on the IHM reports (and counter confirmation from the Yard's own expert HazMat personnel), has the means to properly remove, temporarily store and eventually dispose of hazardous material in a safe and approved manner. Furthermore, that the final disposal by the yard or sub-contractors, fulfils all the requirements of local Governmental Regulations and broadly the requirements stated in the Basel Convention.
- The Yard has the means & proficiency to deal with all foreseeable emergencies.

Furthermore, to ensure the ship or offshore unit to be recycled has an Inventory of Hazardous Materials (IHM) Part I, that such inventory has been approved by Class & is maintained up to date and then, just prior recycling, IHM Part II and III too.

Also, that the ship or offshore unit has received a proper "Ship Recycling Plan" from the Yard which is comprehensive, provides a description of the preparations prior to the ship's arrival, provides a detailed block cutting sequence and flow chart of cutting activity, describes all the safety requirements especially for working in spaces where permits are required, describes the qualifications of personnel involved in special work, specifies the proper removal and disposal of Hazardous Materials and contains the necessary permits required from the relevant authorities.

And that a "Ready for Recycling" certificate - if and when applicable, has been issued by Class or Flag.

5.5 OWNERS ADDITIONAL SAFEGUARDS

Additionally, in order to be comfortable and assured, owners should as far as possible include in their agreements (such as BIMCO Recyclecon) with the recycling yards or cash buyers, requirements that -
Their supervisory representatives will be permitted to attend the recycling site to -

- Check that the recycling process is proceeding according to plan and all human and environmental protective precautions are being observed.
- Ensure that no block cutting is done which may cause debris to fall into the inter-tidal zones or into the waters.
- Diplomatically point out to the yard any shortcomings, ask that they be rectified prior work continuing, and then check that they actually are.
- Report regularly (normally weekly) back to the Owner on the progress and quality of work and on any shortcomings found and corrected.
- Ensure that hazardous materials are disposed of in a manner that renders them non-harmful to the surrounding environment. Also visit the downstream disposal sites/landfills where possible to ascertain compliance with specifications stated in local governmental regulations and Basel Convention guidelines.

On completion of recycling have the representative counter-confirm that the completion certificate is indeed well founded and the job satisfactorily completed.

Figure 4 - Bow Thruster being removed from the hull of a ship being recycled under CTI-MARITEC supervision.

5.5(a) SUGGESTED AREAS FOR INDEPENDENT SUPERVISORS TO CONCENTRATE ON

In general, it has been found that certified and reputed recycling yards have good standards of safe operations. Nevertheless, described below are a few suggestions for areas that independent supervisors may concentrate on (in liaison with the Yard's Safety department personnel) when overseeing recycling operations in yards in various geographical locations, so as to confirm compliance with safety, health and environmental regulations and also offer practical everyday advice for attempting to achieve excellence in operational standards.

- Work permits (for hot work, cold work, enclosed spaces and working at heights) are properly issued after detailed checks and their validity limits (generally 8 hours) are not exceeded.
- Workers are wearing PPE commensurate for the work they are to perform. (Concentrate on third party workers performing temporary work in the yard)
- Sections containing hazardous materials are prominently marked as such.
- Debris from blocks being cut does not fall into the intertidal zones.

- Ships that are landed in the intertidal zone are hauled in regularly and their double bottom structures kept intact for as long as possible to act as 'impermeable' floors, so as to prevent any pollutive material from escaping into the waters.
- Dismantled blocks kept away from access routes to the ships that may have to be used by emergency vehicles.
- Cranes used for lifting blocks and machinery are operating within their safe working load limits and hauling-in winch wires do not exceed their design loads.
- Adequate pollution response equipment is always kept ready when fuels or other oils are being transferred from the ship to shore.
- Fire fighting equipment such as extinguishers are within the serviced dates.
- Sufficient lighting (natural or artificial) at all workplaces.
- Fuel tanks are properly cleaned and mopped dry prior to any work commencing in them.

Figure 5 – Fuel tank being thoroughly mopped clean

- Areas under machinery are cleaned of residues prior work commences on the machinery removal (example Main Engine sump)
- Cutting gas portable rubber hoses are not led over sharp edges and their connections are free of leaks.
- Dangerous trip or fall hazards are covered and prominently barricaded as well.
- Emergency access/exits to and from the vessel are kept in good order.

6. THE HAZMAT FACTOR IN THE LIFECYCLE OF SHIPS

6.1 FOCUS ON ASBESTOS

As stated before, asbestos is banned on all ships built after 2002. Some exemptions existed on ships built between 2002 and 2011 – for example asbestos was allowed in certain rotary vane compressors & thermal insulation >1000C; however, from 2011 asbestos has been banned entirely.

6.2 ASBESTOS NON-COMPLIANCES

Although new building ships are delivered with an "asbestos free declaration", in many cases asbestos has been found on board during subsequent surveys, or port state inspections. This puts the shipowner in a difficult position as it may sometimes lead to fines, detentions and of course, costs for removal.

6.3 ASBESTOS RELATED FINDINGS OF INTEREST

CTI has been involved in survey testing for asbestos on new builds in which asbestos is forbidden and yet, between years 2011 & 2020, found asbestos in 297 out of 557 new built ships where tests were conducted (53%).

Furthermore, tests on 1824 existing vessels in service showed asbestos present in 1011 of them (55%)

All results obtained from careful analysis of samples at specialist laboratories.

Locations in which asbestos was found (percentage wise) are as follows –

ASBESTOS FOUND IN	RATIO
Boilers	4.13%
Heat exchangers	5.07%
Exhaust Gas Economisers	1.10%
Machinery - Pumps/Compressors/Purifiers	25.50%
Inert Gas Systems	0.07%
Fuel, Hot Water, Condenser Tank cladding	8.51%
Incinerators	1.39%
Piping flange gaskets	26.05%
Valves	12.12%
Diesel Engine structures	5.07%
Pipe Ducting	2.27%
Propeller shafting associated material	0.84%
Air Condition Systems	0.53%
Ceiling and bulkhead insulation	0.96%
Sealing putty	0.17%
Electrical cable sheathing	0.10%
Door and Porthole packing	0.05%
Deck Coverings	0.05%

Table 3- Location of asbestos found on ships supposed to be 'asbestos free'

Figures shown above regarding asbestos, especially on new builds, are of great concern to the maritime industry and must be taken cognizance of by those involved in the ship design and ship building industry.

Figure 6 – CTI-MARITEC Chemistry Laboratory

6.4 CONSIDERATIONS FOR OTHER HAZMAT

A few considerations for reduction of other hazardous material on ships would be –

- Fitting smoke detectors that are of the optical type rather than radioactive type
- Fitting non lead-acid batteries in battery banks, although prices would be a factor to be considered
- Fitting equipment such as transformers that do not use Polychlorinated Bi-phenyls (PCB) as a cooling medium and fitting electric cables that are PCB free.
- Avoiding the use of equipment such as level gauges, thermometers, lamps, fire detectors, gyro compasses that use Mercury.

7. CONCLUSIONS

In the world today the betterment of living conditions and health of human beings is being given increasing credence, mainly through the protection of nature and the environment. Also considered is the need to ensure a better life for future generations.

As such, reduction of hazardous materials in the construction of ships so as to make them safer for crew who sail on them, make it safer for workers who dismantle them and to reduce the need to dispose of tonnes of hazardous materials in the future, is a factor that must be considered.

It is hoped that the three topics covered above would cause readers to understand why the Regulations that govern the Inventorying, Handling and Disposal of Hazardous Material are therefore necessary & need to be followed as best as possible by in the Maritime Industry.

Also that the design of future ships will take into consideration the need to minimize Hazardous Materials onboard – mainly the responsibility of Owners when making specifications and those who build ships - Naval Architects and Ship Construction Yards.

8. AUTHOR'S BIOGRAPHY

John Ren Di currently holds the position of General Manager at CTI-MARITEC Environmental Services, Singapore. He is responsible for the provision of multiple services offered by CTI-MARITEC to the Maritime Industry including Inventory of Hazardous Materials on ships, Maintenance of the Inventory, Responsible Ship Recycling Supervision, Ballast Water System Commission Testing, and also Fuel Quality Testing and Analysis. John started his Maritime Career as a Port State Control Officer in the 1990s. In 2010 he switched to conducting Marine Hazardous Material (HAZMAT) Surveys and ever since has been heavily involved in HAZMAT management of new building and existing ships, also of ships bound for recycling. He is also, amongst other things, the Technical Representative of the Singapore Ship Owners Association in the field of ship recycling.

IMPROVING ENVIRONMENTAL IMPACT AND REFINING FLEET RELIABILITY THROUGH THE USE A RELIABILITY CENTRED MAINTENANCE PROCESS TO IMPROVE MAINTENANCE OF THE VESSEL FLEET OF THE MALDIVES TRANSPORT AND CONTRACTING COMPANY PLC

I Ridley, University of Sunderland, UK

In the current world economic situation, fleet operators are striving more than ever to ensure their fleet delivers the best possible return. An important aspect of delivering the best possible return is by ensuring the highest possible fleet availability.

A prime mover in ensuring fleet availability is an up to date, appropriate and cost effective maintenance regime. Such a maintenance regime can help to ensure that vessel down time is kept to a minimum so as to ensure maximum revenue generation.

Utilising a Reliability Centred Maintenance Process to review an organisation's various maintenance regimes and breakdowns/failures will help to ensure that wherever possible the most appropriate and cost effective maintenance regimes are applied.

This paper looks at aspects of the Reliability Centred Maintenance Process and then explores the challenges of applying such a process to an actual fleet of vessels. It discusses how appropriate data is obtained and then how that data can be used to develop an improved and appropriate maintenance regime throughout a fleet of varying vessel types.

This study is based on work carried out by the Maldives Transport and Contracting Company Plc. (MTCC) and how Reliability Centred Maintenance Process could be used to improve availability and reliability of the varied fleet of ferries and other vessels. The ferry services provided by MTCC connects 85% of the population of the Maldives and so it is vital service to the majority of the country's population and hence the importance of this study.

The Royal Institution of Naval Architects
8-9 Northumberland Street
London, WC2N 5DA
United Kingdom

ISBN 978-1-7138-8305-0

CURRAN ASSOCIATES INC.
proceedings.com

2nd Workshop on Natural Language Reasoning and Structured Explanations

Bangkok, Thailand
15 August 2024